Thomas C. Lothian Pty Ltd
132 Albert Road, South Melbourne, Victoria 3205
www.lothian.com.au

National Library of Australia
Cataloguing-in-Publication data:

Crew, Gary, 1947- .
I did nothing : the extinction of the gastric-brooding frog.

For younger readers.
ISBN 0 7344 0507 3.

1. Frogs - Juvenile fiction. 2. Endangered species - Juvenile fiction. I. Wilson, Mark, 1949- .
II. Title. (Series : The extinct series).

A823.3

Design by John van Loon
Colour reproduction by Prism Graphic Services
Printed in Singapore by Imago Productions

I Did Nothing

The Extinction of the Gastric-Brooding Frog

Gary Crew & Mark Wilson

Lothian
BOOKS

ZEEK put the package on the dissecting table and backed away.

'What's that?' I wanted to know.

'Dunno,' he said, making a dopey face. 'Came in this morning's mail. It's addressed to you.'

Zeek had just finished uni. His fingernails were so clean they shone. Worse, there wasn't even a stain on his lab coat.

'Well, open it,' I said. 'I'm busy …'

The package was wrapped in brown paper and tied up with string. Julie Andrews would have loved it. Zeek picked up a scalpel, cut the string and unwrapped the paper. There was a cardboard box inside. When he lifted the lid, I saw the body of a frog packed in scrunched-up newspaper. 'Phew!' I sniffed. 'It's been preserved in metho!'

'Weird,' Zeek muttered. And seeing there was no covering letter, he turned the wrapping paper over. 'Here,' he said. 'On the back. See? Sender: Cory Wells, Days Road, Conondale. Queensland. 4552.'

'Conondale!' I said. 'You know what this is?'

'A dead frog … ?' He made the dopey face.

I sighed. I was sick of trainees. 'What sort of frog?'

Zeek looked vague.

'The sender's address is Conondale, right?' I prompted. 'And look here … It's got eyes on the top of its head …'

'Dorsal protrusion,' he said.

'Good. So … ?'

ZEEK didn't answer right away. Then he let fly. 'This is what's left of a gastric-brooding frog: *Rheobatrachus silus*. Only found in creeks and ponds in the rainforest around Conondale. Caused a sensation when it was discovered in 1973. Carried fertilised eggs in its stomach and gave birth from its mouth. It could close down its acidic gastric juices and turn its stomach into a womb. Could have proved a real bonus treating stomach problems in humans. Maybe even cancer. But …'

I was impressed. 'But what?'

'But … it's supposed to be extinct!'

'Supposed to be,' I said.

'Well, it hasn't been seen in the wild since 1981. That's eight years. All it took for humans to wipe it out — even though it was one of the most highly evolved amphibians that ever lived. Awful …'

'More like tragic,' I said. 'But why send it to me?'

He gave me a look. 'That's not so hard. Anyone who saw your TV special would know that you're the museum's expert on frogs, right?'

'Right, but a live specimen would have been better.'

'If there was one.'

I was warming to this boy. 'Aren't you overlooking something?' I said.

'What?'

'If this frog died last month — even last year — there might still be others where it came from. No animal is said to be officially extinct until fifty years after its last sighting. That's not until 2031. So, you never know.'

Zeek was about to make the dopey face, but I biffed him. 'Well, what are we going to do about it?'

Hard question, he was thinking. *Real hard question. Gotta get it right.* 'Dissect it?' he said.

'No, we're not going to dissect it. At least, not yet … We're going to pay Cory Wells a visit.'

Cory Wells
Notes for Biology Assignment, Term II

1.

2.

3.

4.

These four pictures show the stages of birth of Rheobatrachus silus.
(teachers photo-copies)
It all happens in the mother's stomach.

"...what's left of a Gastric Brooding frog. Rheobatrachus silus,' he said. 'It ...ried fertilised eggs in its stomach and gave birth from its mouth. Caused a ...nsation when it was discovered. Very limited habitat. Only found in creeks ...ponds in the rainforest around Conondale and Kenilworth. Discovered

This was in the Conondale News last month.

Mouth birth, weird!

protruding eyes

Great Dividing Range

Eungella National Park

...isbane

Conondale Range

Sydney

gone?

A S it turned out, we couldn't get to Conondale for another twelve months. Governments don't always consider scientific research a funding priority. It might have been different if we were looking for oil, not for something as unimportant as a frog. But Zeek never stopped talking about it all the same.

One day he said, 'Why *do* you reckon Cory Wells sent us that frog?' This was not a bad question, especially since we had ended up dissecting the thing, only to discover that it had been dead for about fifteen years.

'That's what we have to find out,' I told him. 'Maybe the dead one was something like a promise … ?'

W HEN we finally arrived at Conondale, we left the four-wheel
drive and walked towards the weatherboard house that the local
postmistress had informed us belonged to Cory Wells. 'Did you call
this guy and tell him that we were coming?' Zeek asked.

'No,' I said.

'Then he mightn't even be here,' he said.

'That's exactly why I didn't call him, dopey. If he knew we were
coming, he might have done a bunk.'

'Why would he do that?'

'Um … because killing an endangered species is a crime under
the Conservation Act.' I didn't bother to look, but I guessed Zeek
made the face.

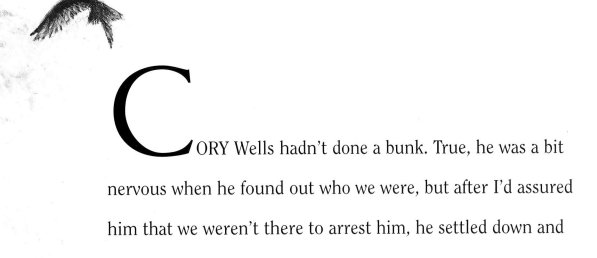

 CORY Wells hadn't done a bunk. True, he was a bit nervous when he found out who we were, but after I'd assured him that we weren't there to arrest him, he settled down and got us a beer.

We sat on his verandah overlooking the forest. The leaves were still dripping after the last downpour. 'So,' I said, kicking back, 'how come you sent us that frog?'

I saw his face fall. He took a swig.

'Yeah, how come?' Zeek chimed in — a bit too eager.

'Take it easy, Zeek,' I told him. 'We're not here to upset the man.'

'Sorry,' Zeek said, and concentrated on his beer.

But Cory was OK. 'Happened when I was in Year Nine, 'bout twelve or thirteen years ago,' he began. 'I used to hang out with the Gibson boys. Bad lot, they were. Real bad … but I didn't have no choice, eh? Livin' way out here.'

'No,' I said, looking out at the forest. 'You probably didn't.'

'Anyway, after school, we used to go down the creek to fool around. Have a smoke, like …' He swung a brown, muscly arm in the direction of the trees. 'Sometimes we'd take our rifles. We had 22s. We might bag a carpet snake. Pythons, they call 'em these days …'

Zeek passed wind, by way of protest. I think.

'Then one day Nigel Gibson spots these frogs …'

'Gastric brooders?' Zeek asked, leaning forward.

'Didn't know that then,' Cory admitted. I gave Zeek the eye to shut him up. He made the dopey face. Cory went on. 'Funny-lookin' things them frogs were. Eyes on toppa their heads …'

'Dorsal protrusion,' Zeek said. I kicked his boot. He got the message.

'They were lively little buggers. Slippery too. So we ended up makin' nets outta hessian so we could catch 'em.'

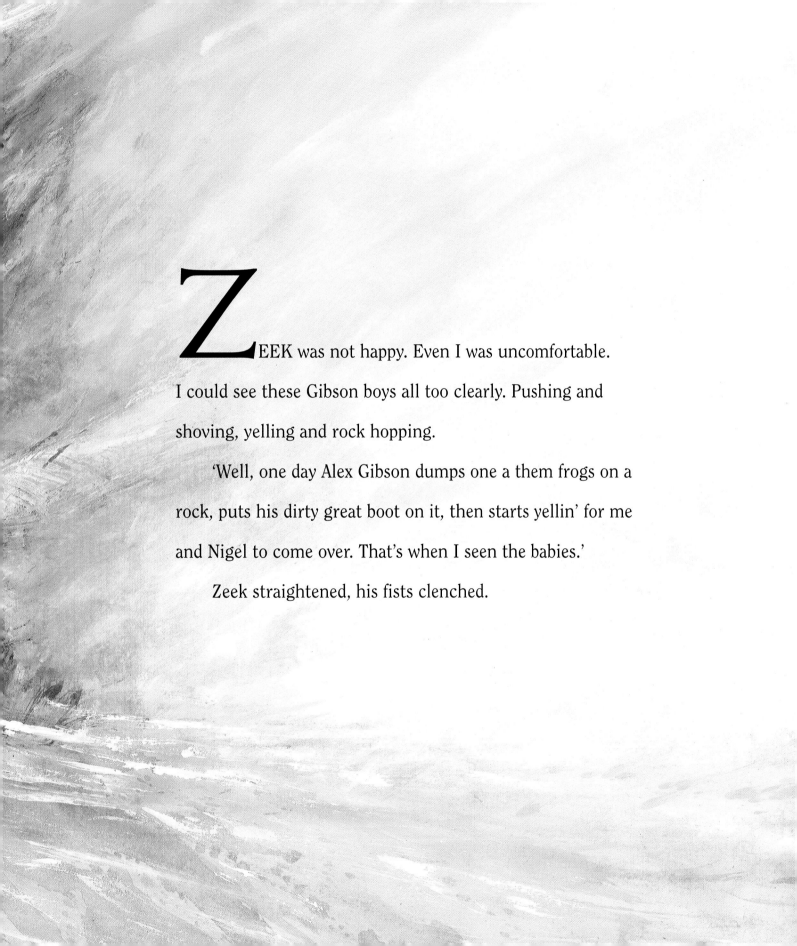

ZEEK was not happy. Even I was uncomfortable. I could see these Gibson boys all too clearly. Pushing and shoving, yelling and rock hopping.

'Well, one day Alex Gibson dumps one a them frogs on a rock, puts his dirty great boot on it, then starts yellin' for me and Nigel to come over. That's when I seen the babies.'

Zeek straightened, his fists clenched.

Cory kept on. ' "Gibbo," I said, "that's a female. It's got babies in its mouth. You oughta let it go." Gibbo didn't. He kept his boot on it then pulled out his pen knife and scraped the babies out. Real tiny, they were. Too little to live, I reckon. But I didn't have the guts to do nothin'. Gibbo was a big guy, so I wasn't game. Them frog babies got washed over the falls. I reckon they died, eh?'

'Probably,' I said. 'What about the mother?'

'Well, since Nigel and Alex was so busy chasing the babies, they forgot about her, didn't they? So I took her up the creek and let her go. Then I went home. I'd had enough of those murderin' Gibsons.'

'Yeah,' Zeek said. 'You sure told them, eh?'

CORY ignored him. 'Next mornin' before school, I went back to check, but the mother was dead. Its legs and arms was stickin' out, like dead frogs do, eh? So I brung it home and stuck it in a jar of metho. I was gunna take it to school, but …'

'But what?'

'I forgot. Didn't think nuthin' more about it until I seen you on that TV show. A fluke, that was. Made me feel bad. I mean, the jar was still on the shelf in the shed. After all those years. All dried out the metho was — an' the frog too, a course — so I put it in a box and posted it to you. Cost me five bucks, it did. Anyway, all that was a long time ago …'

'Sure,' I said.

'I was just a kid, eh?'

'Doesn't matter how long ago it was,' Zeek said. 'You still see kids with nets. And those plastic bug catchers …' He slumped in his chair, his half-finished beer in his hand. He was staring out into that wild, green forest. He wasn't making the dopey face now. 'People kill everything,' he muttered. 'For no reason. Terrible …' Then he sat up and turned to Cory. 'You seen any of those gastric brooders since?'

'Nah, never. The loggers ripped the guts outta the forest years ago. Mucked up the whole creek, they did. Runs brown now. If it's runnin' at all …'

WE set up our tent in the National Park camping grounds and went down to the creek for a look-see. We searched for a week, but found nothing.

'I reckon they're gone,' I said to Zeek.

'You don't know that,' he shot back. 'Not for sure.'

'No, they're gone,' I said. 'Fifty years or not, now that I've looked for myself, I'm certain they are. Yep. Extinct.'

Zeek stood next to me at the top of the falls, the forest canopy stretched out below us. 'Maybe they are,' he said, 'and maybe they're not. And maybe that Cory Wells was too scared to do anything, but I'm not. No …' He turned to look out over the forest, his dopey face gone forever. 'I'm young. And I'm smart. So I *can* do something. And I will.'

EPILOGUE

The extinction of the gastric-brooding frog is sobering evidence of just how quickly humankind can destroy an animal species. The frog was known to Science for barely eight years. Even the slightest change in an animal's habitat — whether from land clearing or introduced animals — can change an entire ecosystem, causing extinction.

The extinction of the gastric-brooding frog should be a source of concern for all humans. Apart from being a truly remarkable animal, the frog performed a vital function in its own ecosystem. Not only that, its unique ability to 'close down' its acidic gastric juices and turn its stomach into a womb was never fully understood prior to its extinction. The very humans who brought about the animal's destruction will now never know the benefits it could have offered medical science — and therefore humanity as a whole.

Each of us, in his or her own way, can do something about the terrible loss of animal species worldwide.